Hollyhock plant

Primroses

Passionflower

Hawaiian flower greeting

Bird-of-paradise flower

Flowers

Written by
DAVID BURNIE

DK

DORLING KINDERSLEY, INC.
NEW YORK

A DORLING KINDERSLEY BOOK

Project editor Christine Webb **Art editors** Ch'en-Ling, Chris Legee
Senior editor Susan McKeever **Senior art editor** Jacquie Gulliver
Production Catherine Semark **U.S. editor** Charles A. Wills
Photography by Roger Phillips
Editorial consultant David Sutton, The British Museum (Natural History)

First American edition of this Eyewitness™ Explorers book, 1992
10 9 8 7 6 5 4 3
Dorling Kindersley, Inc.,
95 Madison Avenue, New York, NY 10016

Library of Congress Cataloging-in-Publication Data
Burnie, David
Flowers/ by David Burnie – 1st American ed.
p. cm. – (Eyewitness Explorers)
Includes index.
Summary: Describes the physical characteristics and life cycles of flowers and examines kinds of garden
flowers, woodland flowers, desert flowers, and others.
ISBN 1-56458-023-7
ISBN 1-56458-024-5 (lib. bdg.)
1. Flowers – Juvenile literature. 2. Wild flowers – Juvenile literature. 3. Flowers – Identification –
Juvenile literature. 4. Wild flowers – Identification – Juvenile literature. [1. Flowers.] I. Title. II. Series.
QK49.B92 1992   91-58209
582.13 – dc20   CIP
                                                                                          AC

Color reproduction by Colourscan, Singapore
Printed in Italy by A. Mondadori Editore, Verona

Contents

Looking at flowers

It's hard to imagine a world without flowers. Plants that have flowers grow almost everywhere, from gardens to high mountains. Every flower has a particular shape and color so that it can carry out a special job. By the time you've finished reading this book, you'll know what this job is and how the flower does it.

The pansy is a typical garden flower – big, bold, and bright.

This rose flower has a ring of five petals.

Animal visitors
Many flowers have lots of animal visitors throughout the day or even sometimes at night. But what are these animals doing? Read on, and you'll find out.

What's what
There is much more to a flower than just its petals. Further on, you can discover what else makes up a flower and what all the different parts are there for.

Explorers' equipment

Keep records of flowers that you see with colored pencils and a drawing pad. A magnifying glass is useful for looking at flowers closely; scissors and tweezers will help you to investigate the parts of a flower.

Keeping a record

A notebook is the flower explorer's most important piece of equipment. Use it for drawing flowers – and the plants that you find them on. If you want a lasting reminder of a flower's shape, you can learn how to press it later on in this book.

Drawing flowers helps you to see how they are made.

A cherry is a juicy package that contains a single seed.

✋ *Always ask permission before picking a flower.*

Making seeds

Once a flower has withered, it starts to make seeds. Some seeds are packaged in juicy cases. We call them fruit. Later on, you will see how seeds are scattered and how they turn into new plants.

Flowers in close-up

A good way to find out about flowers is to take one apart. The petals are often big and bright. But if you take the petals off, you will be left with the important parts of a flower – the ones that make the seeds.

Simple flowers

Kingcups and poppies are called simple flowers. Their petals are arranged in a circle. In the center of the flower are the parts that produce the seeds.

Kingcup

Seeds form in pointed green parts, called ovaries.

Center has been cut in half.

Green sepals protect the bud.

Poppy

The yellow tufts around the center of the flower are called anthers.

Petal

Poppies were once common weeds in wheat fields. Whole fields were turned red by their flowers.

Many flowers in one

From a distance, this thistle seems like a single flower. But take a closer look. It is made of many tiny flowers or florets, packed together. In the daisy's "face," you will see lots of tiny dots. Each of these is a tiny flower. Flowers like thistles and daisies are composite flowers.

Thistle

Each tuft is a flower.

Inside the thistle, you can see the tube-like florets.

Tufty thistle

Look around in your garden or in the park, and see how many composite flowers you can spot. The tiny florets of this thistle make a tuft that looks like a brush.

The "face" is made up of many tiny florets.

Daisy pieces

Try to pull apart a large daisy to see the different florets that make it up.

Each "petal" is really a separate, lopsided flower.

Each floret makes a single seed.

11

Complicated flowers

If you look at flowers in a garden or in the countryside, you will see that they come in many shapes. Some flowers are flat and round. But others are shaped like funnels, beaks, or even umbrellas. They are complicated flowers.

The seed-making parts are hidden inside the "beak."

How many petals?

An everlasting pea has five petals, but you need to look closely to see them. Two are joined together to make a "beak" that sticks forward. Two more lie on each side, while the fifth makes a curved "hood" around the top of the flower.

"Beak" made up of two petals

Piped aboard

The Dutchman's-pipe plant has flowers that look just like a pipe. They attract flies with their sickly smell. The flies can escape only when the flower withers.

"Hood" petal

"Beak" petal

Anthers hold a yellow dust called pollen.

"Beak" petal

The hood has a slippery lining.

Spadix

A living trap
Imagine being caught in a trap, dusted with pollen, and then released. That is what happens to tiny flies that visit the lords-and-ladies plant.

Flies tumble from the spadix or the hood into a lower chamber, which contains tiny flowers.

The club-shaped spadix gives off a scent that tiny flies like.

Lower chamber

Try to guess how many flowers a flower head has, and then count them up. You may be surprised by the difference!

Growing together
Many complicated flowers grow together in clusters called flower heads.

13

Flowers with a difference

In nature, things are not always what they seem. Some plants have small, drab flowers, but they still manage to put on a beautiful show of color. Instead of petals, they use brightly colored leaves or sepals to attract insects.

Two in one

Take a look at this hydrangea flower head. It is made up of many flowers tightly packed together. Some of the flowers have big, colorful sepals, while the others are quite small.

This is a "lacecap" hydrangea.

Inner flowers are much smaller. Only these can make seeds.

Bracts

Outer flower has four big, colorful sepals to attract insects.

Find that flower!

Most of this sun-spurge flower head is made up of special cup-shaped leaves called bracts. The flowers themselves are tiny, and they nestle in the middle of each cup.

Tropical treat

You have to look closely to see this plant's true flowers, because they are very small. They are surrounded by special pink leaves. The plant's name, bougainvillea, is quite a mouthful. Try saying *boo-gan-vil-ia*.

Flower

Ordinary leaf

This colored leaf is called a bract.

Sepals

Leaf stems cling on by hooking around other plants.

Showy sepals

The showy "flower" of this clematis is actually a ring of sepals. Many plants have small green sepals, but the clematis has sepals that are bigger and more colorful than the rest of the flower.

Colorful leaves

When growing wild, poinsettia plants (say *poyn-set-ee-ah*) grow into big bushes topped by leaves that turn crimson during the flowering season. The poinsettia's real flowers are surrounded by these bright leaves.

Bougainvillea is a climbing plant.

Plants that don't flower

No matter how well you care for a fern, you will never make it flower. Nor will you ever see flowers on a moss, or a lichen, or any of the plants shown here. This is because these plants have neither flowers nor seeds. Instead, they reproduce by making dust-like spores.

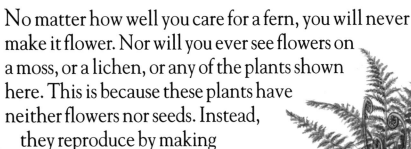

A fern's leaves, or fronds, uncoil as they grow.

Tree ferns have thick "trunks." Some grow as high as a house.

Many ferns carry their spores on the back of fronds.

Feathery ferns
Most ferns live in damp places. Their leaves, or fronds, are often split into many pieces, giving them a feathery outline. An easy way to spot ferns is to look at their new fronds. These usually have a coiled shape, and they unwind as they grow bigger.

Slow but sure

Look carefully at rocks, walls, and tree trunks for lichens. They look like flat patches that have been stuck on. They grow very slowly but live for a long time. One as big as your hand could be much older than you are!

Some lichens are brightly colored.

A lichen expands outward as it grows.

Mushrooms and toadstools

Fungi are not true plants. They feed on the remains of other living things. A mushroom or a toadstool is just the part of a fungus that makes spores. The rest of the fungus is hidden away.

The frilly "gills" of a toadstool drop spores into the air.

Wet and dry

Mosses are small plants that are often shaped like little cushions. They usually live in damp places. But some mosses live on walls and rooftops, where it can get very dry. In hot weather, they turn gray and hard. When it rains again, they turn green and start growing once more.

Slime and seaweed

In summer, ponds sometimes fill with green slime. This is made of tiny algae – very simple plants that have no flowers. There are thousands of kinds of algae, including seaweed and the green "dust" that covers tree trunks.

The life of a flower

Have you ever noticed how wild poppies can suddenly spring up on a patch of bare ground? This happens because most poppies are "annual" plants. They grow very quickly, and they flower and die all within the space of a year. Not all plants are like this. Many more live for a number of years. Plants like this are called "perennials" (say *per-en-ee-als*).

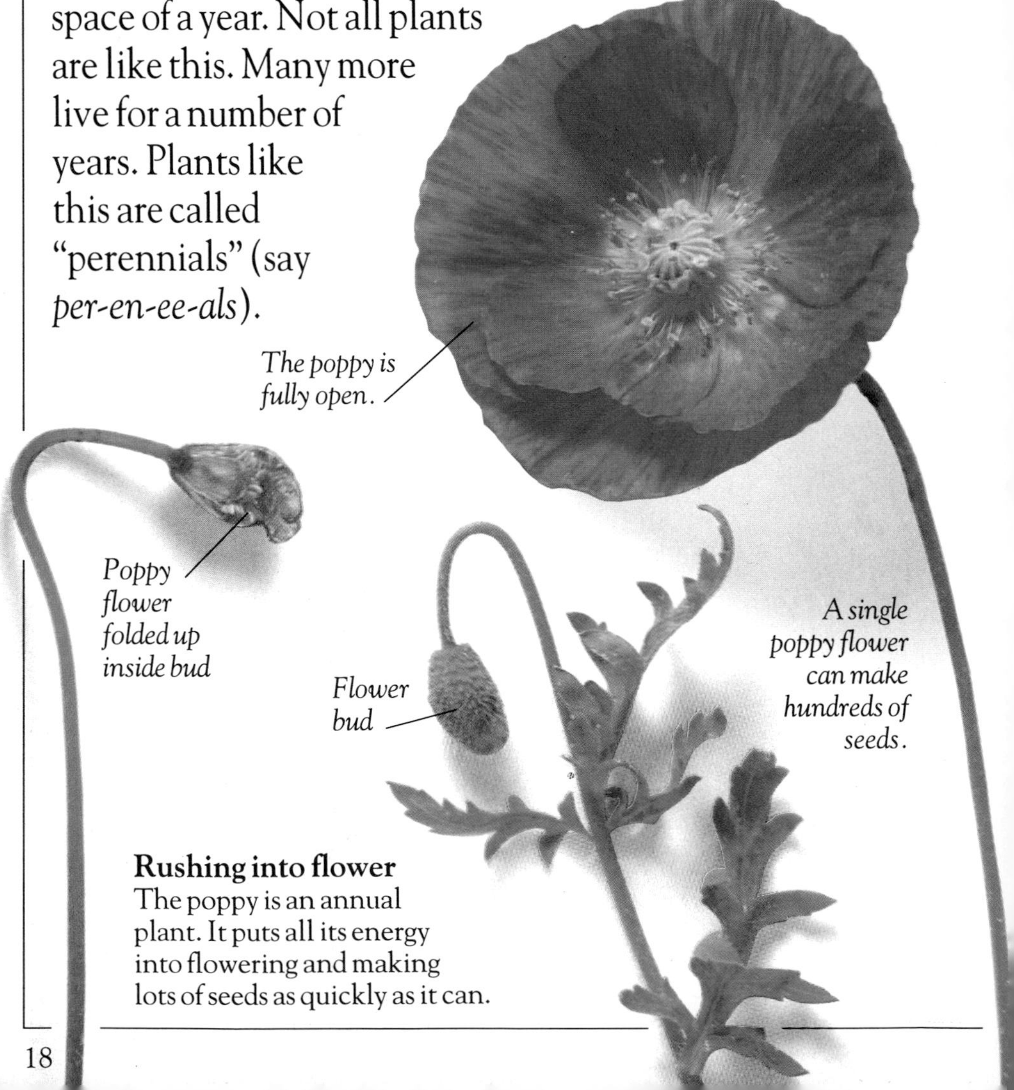

The poppy is fully open.

Poppy flower folded up inside bud

Flower bud

A single poppy flower can make hundreds of seeds.

Rushing into flower
The poppy is an annual plant. It puts all its energy into flowering and making lots of seeds as quickly as it can.

Long-lived plants

These garden cranesbills are perennials.
They have big leaves and spreading roots.
In the autumn, the leaves die but the
roots stay alive through the winter.
In the spring, new leaves grow and
cranesbills flower once again.

*The poppy
petals start to
wither and
fall off.*

*Perennial plants grow
best where the ground
is not disturbed – from
woodlands to deserts.*

*A single
cranesbill
flower makes
just five seeds.*

*Annual plants like this
poppy grow best where the
soil has been dug, plowed,
or moved around.*

*Chamber
containing
seeds*

*Seeds are shaken
out by the wind.
The cycle begins
again the next
year, when the
seeds germinate.*

*Seeds are
made in this
chamber.*

Bursting into flower

A flower bud is like a well-packed suitcase. It has a tough outer cover, which stops it from being damaged. Inside, the different parts of the flower are rolled up tight, so they take up very little space. As the bud grows, the flower expands inside. Soon, the flower becomes so big that it can no longer fit in the bud. Then it bursts into bloom.

Wild yellow irises grow by ponds.

Petals unfold after bursting through the bud's papery sepals.

Iris flower buds grow out from folds in the pointed leaves.

The buds are protected by sepals.

The highest bud flowers first. When it withers, the next bud opens.

The three petals of an iris that point upward are called standards. In this iris, they are frilly and rounded.

Calling all bees
An iris is not just a pretty shape – it is a complicated machine that attracts bees to visit it. Once a bee has landed, the flower sprinkles it with a special dust called pollen.

The three drooping petals are called falls.

Bees use falls as landing platforms.

Streaks on the petals guide bees to a sweet food called nectar at the center of the flower.

Tightly rolled petals

Ovary

Anther coated with pollen

Inside the bud
This iris bud has been sliced open just before it was about to bloom. See if you can make out the rolled-up petals and the ovary, where the seeds are made.

This lily bud is shaped like a torpedo. It has six long, pointed petals.

Blooming colors

Flowers use their colors to attract insects. When a bee or a butterfly sees a brightly colored flower, it flies toward it. The color is like a signpost that shows where tasty nectar can be found.

If you wear a bright shirt on a sunny day, insects may land on you because they think you are a flower.

The colors in petals are made by chemicals called pigments.

Do-it-yourself colors

To see how a flower draws up water, take a white carnation and ask an adult to split the lower half of the stem in two. Put one side in tap water and the other in water with food coloring. Within an hour, one half will change color! This happens because the flower draws up the water and the coloring through the stem.

Color test

Flowers often have many pigments in their petals. To see them, try this experiment. You will need colorful petals, a bowl, a fork, rubbing alcohol, some string, paper clips, and blotting paper.

1 Put the petals in a bowl and mash them into a paste with the fork. Ask an adult to add enough rubbing alcohol to make the mixture runny.

2 Clip a strip of blotting paper over the bowl so that it just touches the mixture. Leave it for an hour in a well-aired place, like on a window ledge. Then check to see what has happened.

Green for growth

Plants have a green pigment in their leaves called chlorophyll. Chlorophyll is very important for plants, because it collects energy from sunlight so that they can grow.

Tie string to two tall objects.

The blotting paper soaks up the liquid, separating it into different bands of color.

Each color that you see is a pigment.

Dyes from plants

Thousands of years ago, the Greeks and the Romans used plant pigments to dye their clothes. They used saffron, from the flowers of a crocus, to make yellow. They made blue colors from the woad plant, and red came from the root of the madder plant.

All about pollen

A plant cannot grow seeds until two kinds of cells join up.
One kind of cell is called an ovule. Ovules are formed in
the base of the flower, and they are kept safe in a chamber
called an ovary. The other kind of cell is called a pollen
grain. Pollen grains have to join up with the ovules from
another flower, so they have to move from
one flower to another.

Stamen

The pollen grains are made at the end of the stamens, in anthers.

Coming and going
This lily flower makes
ovules as well as pollen
grains. But it will only
form seeds if it receives
pollen that has traveled
from another lily flower.

Stigma

The stigma is a landing platform that receives pollen from other flowers.

Pollen grains are tiny because they have to move from one flower to another.

Anther

Stigma

A *pollen grain lands on the stigma.*

The grain grows a slender tube and joins up with an ovule.

A seed is formed here, in the ovary.

Hollyhock plant

This hollyhock pollen grain has been magnified.

Pollen at the ready

This lily bud has been cut in half to reveal its stamens and stigma. At the end of each stamen is a dark orange anther. This makes the pollen, which is like a fine dust.

Stigma

Anther

A single anther makes millions of pollen grains.

Pollen in close-up

Pollen grains are so small that about 50 of them would fit on a pinhead. Every plant has its own type of pollen. Some grains are round, while others are shaped like triangles or sausages.

Sticky business

Pollen grains are often sticky. When a bee visits a flower, it cannot help brushing against the anthers and getting covered with pollen.

Animal visitors

Have you ever noticed insects flitting from flower to flower? They are spreading pollen. When a bee visits a flower, it gets dusted with pollen. When it moves on to another flower, it unloads some of the pollen and picks up some more. In return for its hard work, the bee gets "paid" with sugary nectar.

Geoffroy's bat visits flowers at night.

The petals are tough so that they are not damaged by bats and birds.

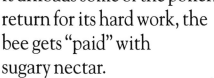

Bright colors attract bats and birds.

Birds and bats
The bird-of-paradise flower is pollinated by birds and by bats. It is big and bright, and produces lots of sweet nectar.

The part of the flower that makes pollen is like a perch. It dusts pollen onto its visitors' feet.

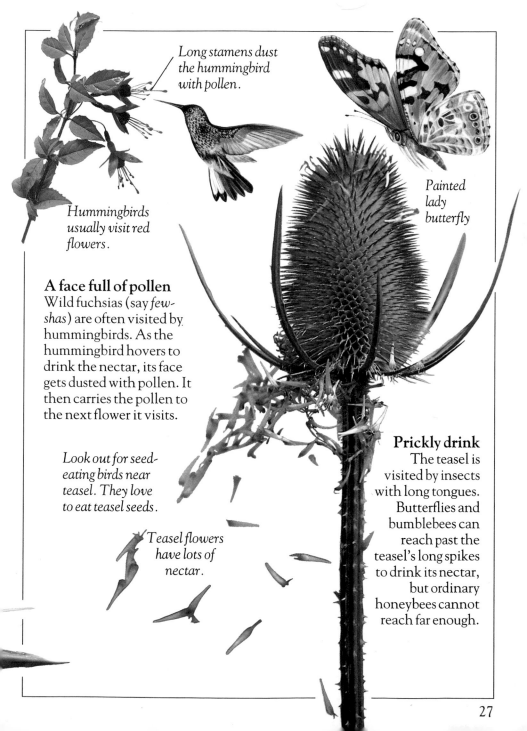

Long stamens dust
the hummingbird
with pollen.

Painted
lady
butterfly

Hummingbirds
usually visit red
flowers.

A face full of pollen

Wild fuchsias (say *few-
shas*) are often visited by
hummingbirds. As the
hummingbird hovers to
drink the nectar, its face
gets dusted with pollen. It
then carries the pollen to
the next flower it visits.

Look out for seed-
eating birds near
teasel. They love
to eat teasel seeds.

Teasel flowers
have lots of
nectar.

Prickly drink

The teasel is
visited by insects
with long tongues.
Butterflies and
bumblebees can
reach past the
teasel's long spikes
to drink its nectar,
but ordinary
honeybees cannot
reach far enough.

27

Perfumed flowers

Have you ever wondered *why* flowers smell? The answer isn't to please our noses. Instead, flowers use smell as a signal. Their scent spreads into the air, where bees and other insects can detect it. The insects fly to where the scent is strongest. This leads them to the flowers and a sugary meal. Most flowers smell strongest by day, but a few release more of their scent at night.

Freesia flowers open one after the other. Each flower lasts for several days.

Bees are attracted to flowers with a sweet smell and bright colors.

Bell-shaped flowers

Sweet-smelling freesias
Freesia flowers produce a rich scent for many days, which is why they are often cut and brought indoors. Eleven kinds of freesias grow in the wild, but many more varieties have been specially bred by gardeners.

Calling all bees...
Grape hyacinths have small, bell-shaped flowers. They give off a rich scent during the day, attracting bees in early spring. Wild grape hyacinths grow mainly in gardens and parks.

Smelly monster
The world's biggest flower is the giant rafflesia. It grows up to 2 1/2 feet across and attracts flies by smelling like rotting meat!

Moths have long tongues that can reach deep into the honeysuckle's tube-like flowers.

Look out for hawkmoths darting among scented flowers after dark.

A scent in the night
Try smelling a honeysuckle's flowers during the day and then in the evening. You will find that the evening scent is much stronger. Honeysuckle is pollinated mainly by moths. It releases its perfume after dark to attract its nighttime visitors.

The evening primrose has a strong scent and pale color to guide moths toward it.

Night shift
Many flowers open up during the day and close at night. The flowers of evening primrose work the other way around. As dusk falls, they open wide and release their scent to attract moths.

Floating in the wind

Not all flowers are pollinated by animals. Instead, some use the wind. If you suffer from hay fever, you may know all about this. They shed their pollen into the air, and the tiny grains are blown far and wide. Some pollen lands on the ground, some gets in our eyes and noses, and just enough lands on other flowers.

Summer sneezing
People who suffer from hay fever are allergic to pollen. They suffer most in spring and early summer, when grasses release huge amounts of pollen into the air.

Afloat in a boat
A few plants use water to move their pollen. This is Canadian pondweed, a plant that you might see in fish tanks. It has two different kinds of flower, both on long stalks that reach up to the surface. The male flowers scatter their pollen over the water. Each pollen grain drifts about until it pollinates a female flower.

Stand back!
Stinging nettle flowers shed their pollen into the air in a special way. When the pollen grains are ripe, a tiny explosion in each flower shoots the pollen into the air.

If you stand near a nettle, you can sometimes see the flowers exploding.

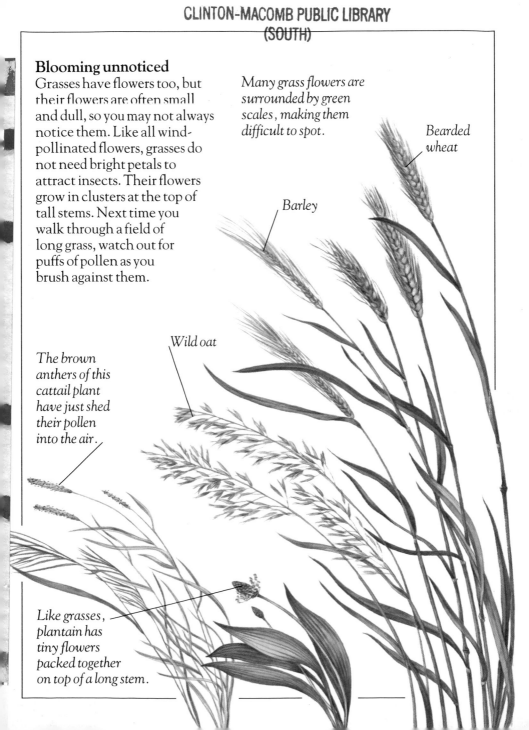

Blooming unnoticed

Grasses have flowers too, but their flowers are often small and dull, so you may not always notice them. Like all wind-pollinated flowers, grasses do not need bright petals to attract insects. Their flowers grow in clusters at the top of tall stems. Next time you walk through a field of long grass, watch out for puffs of pollen as you brush against them.

Many grass flowers are surrounded by green scales, making them difficult to spot.

Bearded wheat

Barley

Wild oat

The brown anthers of this cattail plant have just shed their pollen into the air.

Like grasses, plantain has tiny flowers packed together on top of a long stem.

Seed story

Seeds come in all shapes and sizes. Some are as big as footballs. Others are so small that millions can fit inside a matchbox. When the seeds are ready, they leave the parent plant and begin a new life on their own. Some seeds drop to the ground, but more often they scatter far and wide.

Flyaway seeds
If you blow hard on a dandelion's seeds, they fly away. Each seed has a tiny parachute to help it float in the air. This helps the wind blow them a very long way, so they can take root on a new patch of ground.

Dandelion flower

The baby pick-a-back plant grows out from where the stalk joins the leaf.

No seeds
Some plants don't produce seeds. The baby pick-a-back plant grows from the leaf of the parent plant. When the parent plant bends over and touches the ground, the baby plant takes root.

Seed
head

Flowers
wither and
drop off
when seeds
form.

Fire lovers
Banksias grow in the dry bush of
Australia. Their hard, wooden
seedpods stay tightly shut until a
fire sweeps through the bush.
When the fire is over, the seedpods open up, and
the seeds drop
onto the ground.

Hard pod
containing
a seed

Handle with care
Castor oil seeds
contain a deadly
poison. But
when they are
crushed, the seeds
produce a valuable oil
that is used as a medicine.

Seeds grow
inside a spiny
case.

Burdock
seed head

Castor
oil seeds

Taken for a ride
If you go for a walk in the country, seeds
often stick to your clothes. This is how
burdock seeds spread – by "hitching a
ride" with passing animals and people.

Juicy fruits

Have you ever wondered why some plants pack up their seeds in such a juicy way? The answer is that it helps them to spread. When a bird feeds on berries, it swallows the fruit complete with the seeds. The seeds pass straight through its body and out with its droppings. They land on the ground and sprout, often far from the plant that produced them.

The story of a strawberry
Wild strawberries are small, and it takes a lot to make a mouthful. Garden strawberries have been made bigger by "crossing" different kinds of wild strawberries (pollinating one kind with pollen from another kind) and by growing only the plants that give the biggest, sweetest fruit.

Strawberry flower

A strawberry's seeds are on the outside.

Unripe strawberry

Seed has a hard coat

The red color shows that the fruit is ripe.

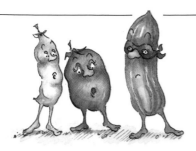

Fruit or vegetable?
Here is a riddle to try on your friends. What do a tomato, a cucumber, and a pod full of peas have in common? The answer is that, strictly speaking, they are all fruits because they all contain seeds.

Known by name
When you go to buy some apples, you will often be asked what kind you want. Apples have been grown for so long that there are now many different varieties to choose from. Each variety has its own name. This apple is a Red Delicious.

An apple's seeds are surrounded by the thick, juicy flesh.

Seeds

Apple blossom
After an apple flower is pollinated, the base of the flower swells up around the seeds to make the apple.

The fig's hidden secret
No matter how much you search, you will never see a blossom on a fig tree. This is because each fig is filled with hundreds of tiny flowers.

Flowers and seeds are hidden inside the fig.

Figs are pollinated by a tiny female wasp. It squeezes through a hole in the end of the fig and spreads pollen over the flowers inside.

Wasp enters here.

Flowers from bulbs

Next time you see someone slicing open an onion, ask if you can have a closer look. (This might make your eyes water!) You will see that it is made up of many tightly packed layers. Onions are bulbs, just like daffodils and tulips. Each layer of a bulb is a store for food. This food is used up when the plant starts to grow.

Parrot tulip

Fit for a prince
Wild tulips grow in hot countries such as Turkey. Hundreds of years ago, Turkish princes grew them around their palaces. Today, many kinds of tulips are grown all over the world.

Daffodil

Flowers that grow from bulbs come in all sizes, shapes, and colors.

Narcissus

Fritillary

Hyacinth

This parrot tulip's frilly, curled petals look like a parrot's feathers.

When you go for a walk in spring, see if you can guess which flowers grow from bulbs.

Crocus

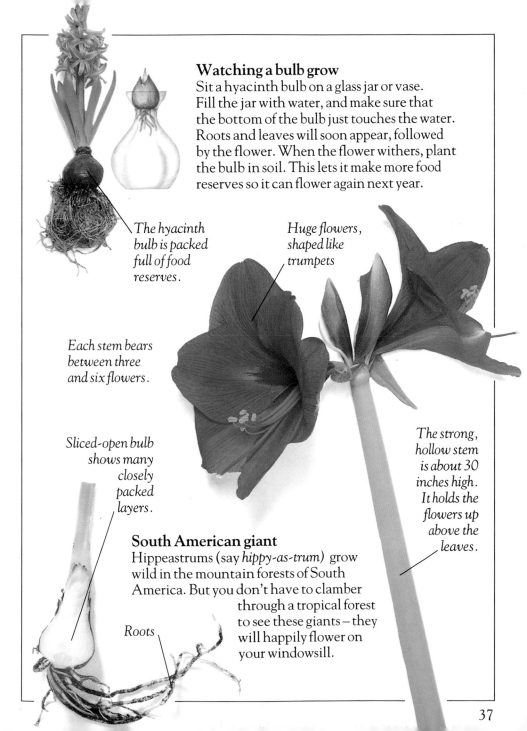

Watching a bulb grow

Sit a hyacinth bulb on a glass jar or vase.
Fill the jar with water, and make sure that
the bottom of the bulb just touches the water.
Roots and leaves will soon appear, followed
by the flower. When the flower withers, plant
the bulb in soil. This lets it make more food
reserves so it can flower again next year.

*The hyacinth
bulb is packed
full of food
reserves.*

*Huge flowers,
shaped like
trumpets*

*Each stem bears
between three
and six flowers.*

*The strong,
hollow stem
is about 30
inches high.
It holds the
flowers up
above the
leaves.*

*Sliced-open bulb
shows many
closely
packed
layers.*

South American giant

Hippeastrums (say *hippy-as-trum*) grow
wild in the mountain forests of South
America. But you don't have to clamber
through a tropical forest
to see these giants – they
will happily flower on
your windowsill.

Roots

Springing into life

Seeds may look dry and dead, but inside each one are living cells waiting for the chance to divide and grow. The wait may be a long one, from weeks to months or years. But as soon as it is damp enough and warm enough, the seed's cells start to divide, and a new plant comes to life. This is called germination.

Seed case falls off to reveal two rounded "seed leaves."

Seed

Green stem appears.

Stem becomes longer.

First steps
Like most seeds, the sunflower germinates root-first so that it can take in water from the soil. The stem then appears at the other end, and as it grows longer, it lifts the seed's hard case off the ground. Eventually, the case splits and falls off.

Root grows downward.

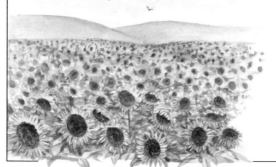

Striking oil
About one-third of a sunflower seed's weight is made up of oil. You may have seen sunflower oil in your kitchen – it is often used in cooking and for making margarine.

A treat for the birds

Most sunflowers are annual plants, which means that they germinate, flower, and die in the same year. Although they live for just a few months, they can get very big indeed. If you grow a sunflower, don't cut it down after it has finished flowering. Instead, leave it and watch the birds arrive to feast on its seeds.

A sunflower's face always points toward the sun.

Sunflowers may grow to more than 10 feet high.

If you plant seeds of a giant variety of sunflower, your plant may grow taller than you!

39

A flying start

A single plant can make hundreds or even thousands of seeds. Each one contains a supply of food. When a seed germinates, the young plant uses this food store to help it through its first few days of life.

Racing ahead

Scarlet runner bean seeds are fun to grow because they get off to a flying start. This is because each fat seed has a large food store packed into two special "seed leaves." In some plants, such as the sunflower, the seed leaves open out above the ground. But in the scarlet runner bean, they stay below the surface.

Stem grows longer and leaves get bigger.

See how Jack's beanstalk grew!

Shoot straightens out above ground.

Hooked shoot grows upward through the ground.

Root grows out of split in seed case.

First leaves open.

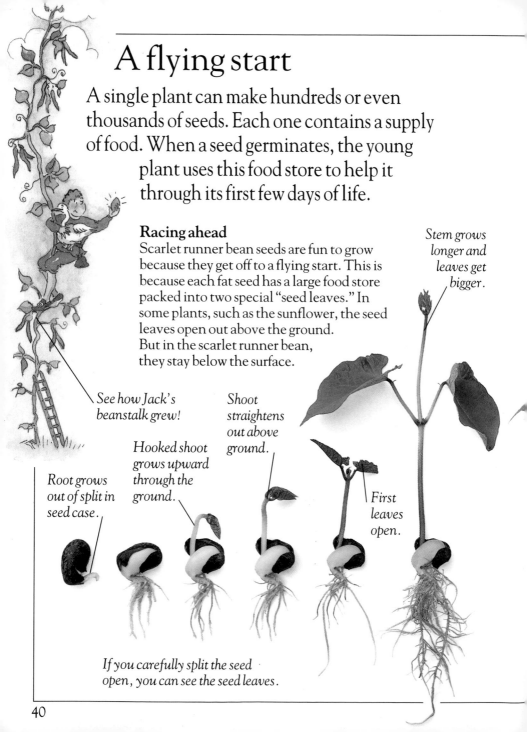

If you carefully split the seed open, you can see the seed leaves.

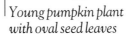

Changing leaves

Seed leaves that open above ground, like the pumpkin's, are easy to spot. They are the first leaves to appear, and they often look different from the "grown-up" leaves that follow.

Young pumpkin plant with oval seed leaves

Second "true" leaf appears.

Second pair of leaves opens out.

First "true" leaf on a bristly stalk.

Seed leaves shrink as their food supply is used up.

Amazing bean!

When a plant germinates, it grows toward the light. You can see this for yourself by growing a bean in a shoebox maze. Cut a window in one end of the box and then make two cardboard shelves that stick out from the sides. Stand the box on end. Now plant a bean in a pot, put the pot inside, and put the lid tightly on the box. When the bean germinates, it will find its way through the maze and out the window! Remember to water the bean as it grows.

Shelf *Window*

You will need scissors, tape, glue, a plant pot, soil, and a bean.

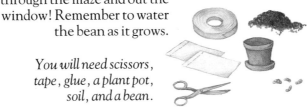

Garden flowers

Have you ever noticed that some garden flowers look very much like wild ones but are bigger and brighter? The reason for this is that all garden plants originally came from the wild. They look different now because gardeners, not nature, decide which plants to breed.

These polyanthuses have many distant ancestors. One of them is the primrose.

Look for wild primroses growing in rows of shrubs and at the edge of woodlands.

Putting on a show
Like many garden plants, these polyanthuses are hybrids. This means that their wild ancestors were not polyanthuses at all, but very different plants. The wild primrose is just one of the polyanthus's distant ancestors.

Garden roses often have many petals.

Wild roses have just five petals.

The War of the Roses

In the English "War of the Roses" more than 500 hundred years ago, both sides used a rose as their badge – one used red and the other white.

Sowing themselves

Wild poppies live all over the world, from the Arctic to the tropics. Garden poppies are easy to grow, and new plants often spring up year after year.

The queen of flowers

Roses have been grown for thousands of years. Most garden roses are much bigger than their wild relatives. They can keep flowering for months, rather than just a few weeks.

Garden pansy

Garden poppy

Wild poppy

Wild pansies

Flowers with a face

Wild and garden pansies both have flat flowers that look like faces. If you look closely, you'll see that the face is made up of five petals. The garden pansy has bigger petals that overlap.

Making flowers last

Most flowers last for a few days, and some last just a single morning. But by pressing or drying flowers, you can make them last far longer. Pressing a flower keeps the outline of its shape, while drying it helps to keep its scent.

Pressing works best with small or papery flowers, such as poppies and pansies.

Pressing flowers

Put a sheet of thick blotting paper on a flat surface, such as a table. Spread the flowers out on the paper. Then cover them with another sheet. Now stack some heavy books on top. Wait at least a week before carefully peeling away the paper to see the flowers.

Using pressed flowers

Pressed flowers are flat, so you can stick them onto cards and gift tags. Remember that they are very fragile, so you must handle them with care. Use a tiny amount of glue – too much will make flowers soggy!

Dried flowers glued to gift tags

Make a flower calendar by drying each month's flowers.

Flowers for all occasions

People often use flowers to mark special occasions. In Hawaii, a flower garland or *lei* is a way of saying "welcome."

Making a potpourri

A potpourri (say *po-pour-ee*) is a colorful mixture of small, scented flowers and petals. To make a potpourri, you will need a tray and some flowers. Pick the flowers when the weather is warm and dry. Leave the small flowers whole but pull the petals off the large flowers one by one. Lay the flowers and petals on the tray, spreading them out to make a thin layer. Every day, stir them around so that the air dries them evenly. The mixture will be ready if it rustles when you stir it. This usually takes at least a week.

Try hanging roses upside down to dry.

Drying flowers

Some flowers will keep their color and scent if you hang them upside down to dry in an airy place.

Put the finished mixture in a bowl, so that the potpourri fills a room with its scent.

Use flowers with a strong scent, such as roses, for your potpourri.

Woodland flowers

All plants need light in order to grow. But woods are often very dark. So how do woodland flowers manage to survive? By keeping one step ahead of the trees. In forests where the trees shed their leaves every year, many plants grow and flower before the trees come into leaf. The plants make the most of the light before the trees shut it out.

Three by three
The wake-robin is a woodland flower that grows in eastern North America. It has a single stem, three leaves, and three petals. The Indians who once lived in the woods made medicines from its fat underground roots.

Wake-robin is also called trillium.

Three leaves grow out around the stem.

Three sepals are beneath the flower.

Flower has three pointed petals.

Scented lily
Lily-of-the-valley (right) has tiny, bell-shaped flowers that give off a rich scent. In the wild, it grows in dry woodlands.

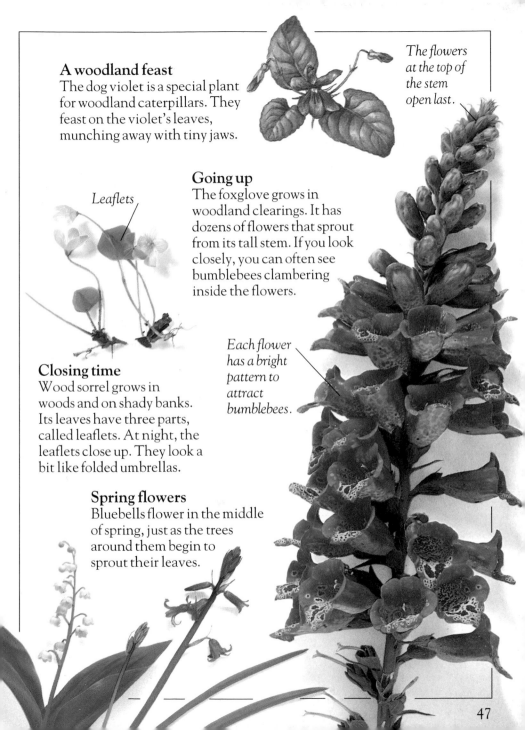

A woodland feast

The dog violet is a special plant for woodland caterpillars. They feast on the violet's leaves, munching away with tiny jaws.

The flowers at the top of the stem open last.

Leaflets

Going up

The foxglove grows in woodland clearings. It has dozens of flowers that sprout from its tall stem. If you look closely, you can often see bumblebees clambering inside the flowers.

Each flower has a bright pattern to attract bumblebees.

Closing time

Wood sorrel grows in woods and on shady banks. Its leaves have three parts, called leaflets. At night, the leaflets close up. They look a bit like folded umbrellas.

Spring flowers

Bluebells flower in the middle of spring, just as the trees around them begin to sprout their leaves.

Tropical flowers

In many parts of the world, plants stop growing during winter. But in the tropics, particularly where it is wet, they can grow and flower all year round. You don't have to go to the tropics to see flowers like these, because in cooler places they are often grown as indoor plants.

Paphiopedilum (paf-ee-o-ped-i-lum) orchids come from Southeast Asia.

Threatened orchids
Orchids grow all over the world, but the biggest and most spectacular live in the tropics. Some kinds have become very rare because too many have been collected and sold.

Insects land on the flower's tail and pollinate it.

The tailflower
Wild tailflowers grow in the forests of South America, but they are popular indoor plants. See how many differently colored tailflowers you can spot.

Shiny leaves stay green all year round.

Brilliant scarlet hood attracts insects.

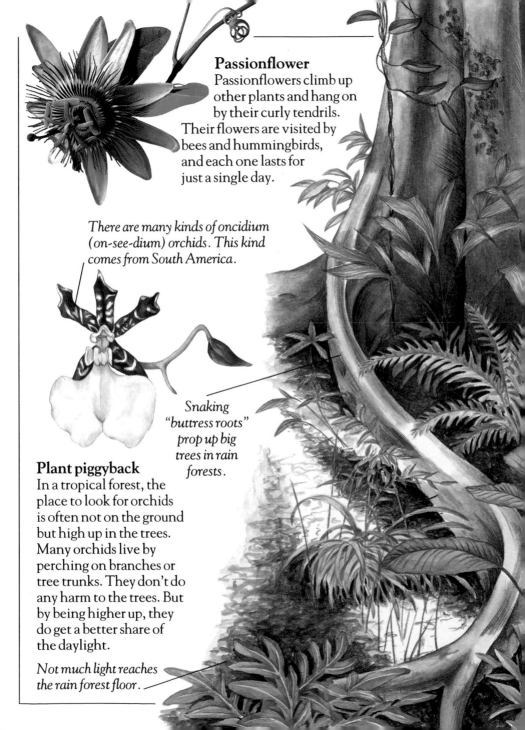

Passionflower

Passionflowers climb up other plants and hang on by their curly tendrils. Their flowers are visited by bees and hummingbirds, and each one lasts for just a single day.

There are many kinds of oncidium (on-see-dium) orchids. This kind comes from South America.

Snaking "buttress roots" prop up big trees in rain forests.

Plant piggyback

In a tropical forest, the place to look for orchids is often not on the ground but high up in the trees. Many orchids live by perching on branches or tree trunks. They don't do any harm to the trees. But by being higher up, they do get a better share of the daylight.

Not much light reaches the rain forest floor.

Grassland flowers

In days gone by, fields and open grassy spaces were often full of wildflowers. But after tractors and weed killers were invented, many wildflowers were plowed up or killed by poisonous sprays. Grassland flowers still survive, but in special places. Look for them in old pastures, on roadsides, around the edges of fields, and on steep banks and slopes.

Harebell flowers grow on long, thin stalks.

Holding its ground

Yarrow is a very tough grassland plant. It can survive being cut by a lawn mower, and it even thrives by the sides of roads.

Nodding bells

The harebell's light blue flowers look like tiny bells as they nod in the wind. Harebell thrives in rough pastures where the soil is shallow.

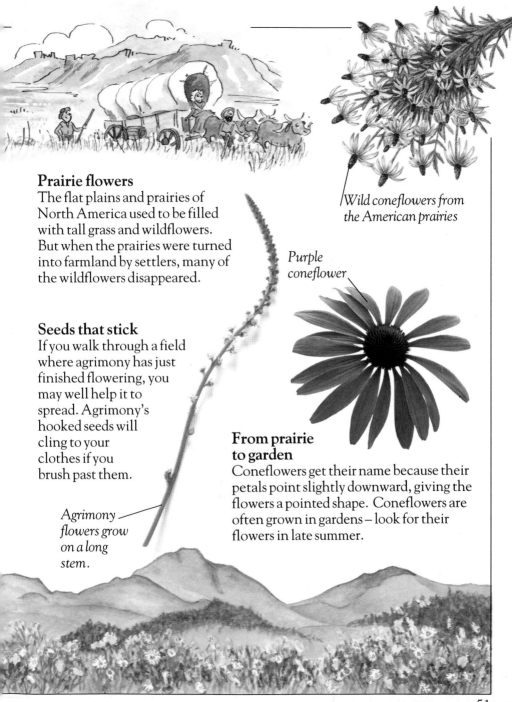

Prairie flowers

The flat plains and prairies of North America used to be filled with tall grass and wildflowers. But when the prairies were turned into farmland by settlers, many of the wildflowers disappeared.

Wild coneflowers from the American prairies

Purple coneflower

Seeds that stick

If you walk through a field where agrimony has just finished flowering, you may well help it to spread. Agrimony's hooked seeds will cling to your clothes if you brush past them.

Agrimony flowers grow on a long stem.

From prairie to garden

Coneflowers get their name because their petals point slightly downward, giving the flowers a pointed shape. Coneflowers are often grown in gardens – look for their flowers in late summer.

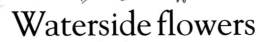

Waterside flowers

The banks of many lakes and streams are like watery jungles – packed with plants and flowers. You'll find that water and marsh plants are choosy about where they grow. Some need ground that is damp but not too wet. Other plants grow in the water but are rooted to the bottom. And a few just float on the surface, with their roots trailing in the water.

Reed-mace (or bulrush) grows in the water. Its flower head looks just like a sausage.

Hemp agrimony grows in damp ground. It has small, tube-shaped flowers in big clusters.

At the water's edge
If you look at the water's edge, you can see how different plants grow in different places. But make sure that an adult is watching when you are exploring near water.

Meadowsweet's creamy-colored flowers have a rich scent.

Monkey flower
Look out for the monkey flower on stream banks. You can see its bright yellow flowers in late summer.

Look for meadowsweet on the stream banks.

Water babies

Some plants grow and flower in shallow water, well away from dry land. In streams, look for water crowfoot. It has long stems that trail in the water's current. On lakes and ponds, look for the large flowers of water lilies and their big floating leaves.

Above the water's surface, water crowfoot has rounded leaves.

Underwater leaves are fine and feathery.

Water lilies have big flowers with fleshy petals. Some have flowers that float on the surface, while others open above the water on a stalk.

Mountain flowers

Imagine you are on a windswept mountainside, with no shelter. You may have to crouch down low to keep warm and stay out of the wind. This is just what many mountain plants do. Instead of growing tall, they are often shaped like cushions or mats, with short stems that are tightly tucked together. This protects them from the wind and helps them to keep warm in the thin mountain air.

Rhododendrons

Rhododendron (say *ro-do-den-dron*) is a long name that just means "red tree." You'll find rhododendron bushes growing in mountain valleys.

Rhododendrons have spectacular trumpet-shaped flowers in late spring.

Thick, shiny leaves

In the Himalaya Mountains in Asia, rhododendrons live as high up as 20,000 feet.

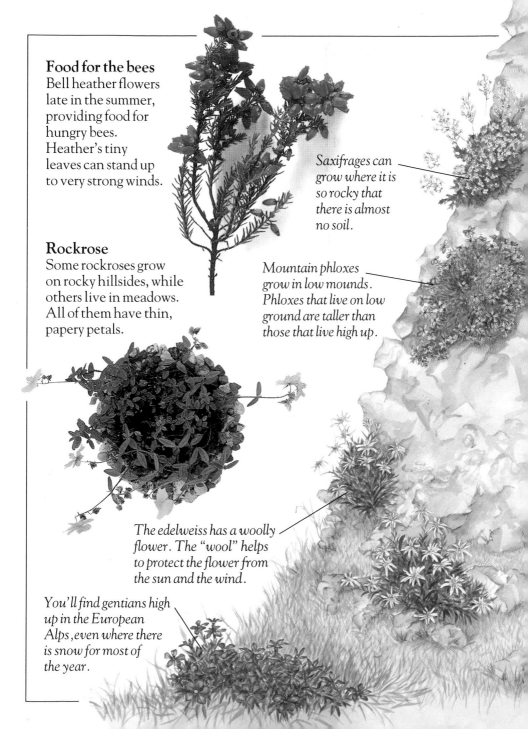

Food for the bees
Bell heather flowers
late in the summer,
providing food for
hungry bees.
Heather's tiny
leaves can stand up
to very strong winds.

*Saxifrages can
grow where it is
so rocky that
there is almost
no soil.*

Rockrose
Some rockroses grow
on rocky hillsides, while
others live in meadows.
All of them have thin,
papery petals.

*Mountain phloxes
grow in low mounds.
Phloxes that live on low
ground are taller than
those that live high up.*

*The edelweiss has a woolly
flower. The "wool" helps
to protect the flower from
the sun and the wind.*

*You'll find gentians high
up in the European
Alps, even where there
is snow for most of
the year.*

Seashore flowers

Plants that grow on the seashore have to cope with bright sunshine and fierce winds, as well as salt from the sea. To do this, most of them have strong stems, tough leaves, and small flowers. Some seashore plants live only on rocky shores, while others live where the shore is made of sand or mud.

Stiff leaves have spiny edges.

Protected by spines

Sea holly grows on rocky beaches and on the edges of sand dunes. Its blue leaves and flowers are protected by hard, sharp spines. Rabbits and other animals eat plants on the seashore, but they leave the prickly sea holly alone.

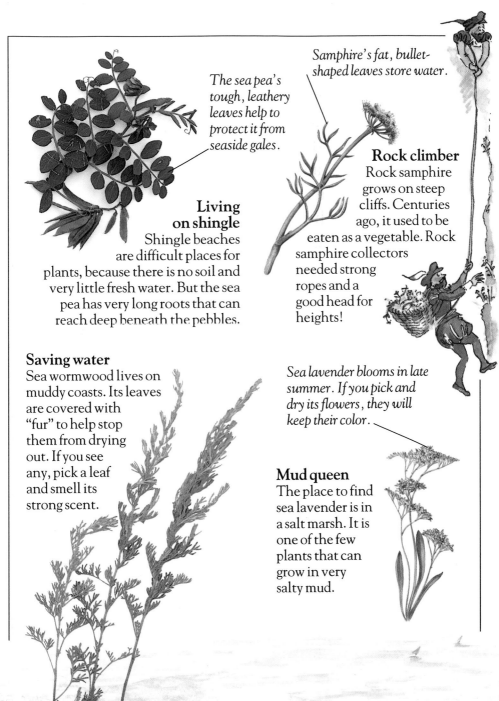

The sea pea's tough, leathery leaves help to protect it from seaside gales.

Samphire's fat, bullet-shaped leaves store water.

Living on shingle

Shingle beaches are difficult places for plants, because there is no soil and very little fresh water. But the sea pea has very long roots that can reach deep beneath the pebbles.

Rock climber

Rock samphire grows on steep cliffs. Centuries ago, it used to be eaten as a vegetable. Rock samphire collectors needed strong ropes and a good head for heights!

Saving water

Sea wormwood lives on muddy coasts. Its leaves are covered with "fur" to help stop them from drying out. If you see any, pick a leaf and smell its strong scent.

Sea lavender blooms in late summer. If you pick and dry its flowers, they will keep their color.

Mud queen

The place to find sea lavender is in a salt marsh. It is one of the few plants that can grow in very salty mud.

Desert flowers

A potted plant that has not been watered is a sorry sight. Its leaves wilt, and eventually it will dry out and die. But some plants can survive for months without a drink. These plants come from deserts – the world's driest places. Many of them have amazing flowers, and they all have special ways of coping with drought.

Each flower has petals shaped like claws.

Flowers of the outback
The Sturt's desert pea is named after one of the first European explorers of the Australian outback. It spreads over the sandy ground to form a low bush.

Leaves have a leathery "skin" that stops them from drying out.

Rotten aroma

The carrion flower lives in dry places in Africa. It has an unusual "perfume" – it smells just like rotting meat. In this way, it attracts meat-eating flies to pollinate it.

Make a mini-desert

Try making a mini-desert in an old aquarium. Add some sand mixed with earth and decorate the surface with stones and dead wood. Then, wearing thick gloves, plant some cacti in the soil. Keep the desert in a sunny spot and water it sparingly in the winter and in spring.

Flower smells like rotting meat

Tough, water-storing stems

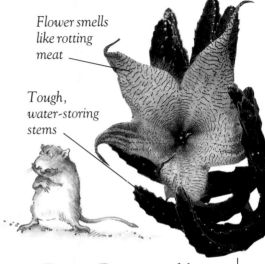

Cacti have star-shaped flowers in spring and summer.

Danger: Do not touch!

You'll find wild prickly cacti in the North and South American deserts. Most of them have swollen stems but no real leaves. Cacti have a store of water that helps them survive long droughts.

Index

A

agrimony, 51
anthers, 10, 12, 21, 24, 25, 31
apple, 35

B

banksia, 33
bean, runner, 40, 41
berries, 34
bird-of-paradise flower, 26
bluebell, 47
bougainvillea, 15
bracts, 14, 15
buds, 10, 18, 20-21, 25
bulbs, 36-37
bulrush, 52
burdock, 33

C

cactus, 59

Wild yellow irises

carnation, 22
carrion flower, 59
cherry, 9
chlorophyll, 23
clematis, 15
colors, 8, 14, 15, 22-23, 26, 28

composite flowers, 11
coneflower, 51
cranesbill, 19
crocus, 23, 36
cucumber, 35

DE

daffodil, 36
daisy, 11
dandelion, 32
Dutchman's pipe, 12

edelweiss, 55
evening primrose, 29
everlasting pea, 12

F

fig, 35
foxglove, 47
freesia, 28
fritillary, 36
fruits, 9, 34-35
fuchsia, 27

Wild coneflowers

GHI

gentian, 55
germination, 19, 38-39, 40-41
grape hyacinth, 28
grasses, 30, 31, 50, 51

harebell, 50
heather, bell, 55
hemp agrimony, 52
hippeastrum, 37

hollyhock, 25
honeysuckle, 29
hyacinth, 36-37
hydrangea, 14

iris, yellow, 20-21

KL

Purple coneflower

kingcup, 10

lily, 21, 24, 25
lily-of-the-valley, 46
lords-and-ladies, 13

MNO

madder, 23
meadowsweet, 52
monkey flower, 52

Rafflesia

narcissus, 36
nectar, 21, 22, 26, 27

onion, 36
orchid, 48, 49
ovary, 10, 21, 24, 25

Wild pansies

RS

Iris

TV

WY

Acknowledgments

Dorling Kindersley would like to thank:
Neil Fletcher for special photography on pages 40-41.
Sharon Grant for design assistance.
Ted Green for collecting flowers for photography.
Gin von Noorden and Kate Raworth for editorial assistance and research.
Jane Parker for the index.
Bob Press for help with authenticating text.

Illustrations by:
Jane Gedye, Keith Grant, Valerie Price, Pete Visscher.

Picture credits
t=top b=bottom c=center
l=left r=right
Bruce Coleman Ltd: 5cr, 29c; /Adrian Davies 27br; / A. Healy 58.
Eric Crighton: 4b.
De Lory: 49tl.
Science Photo Library: / Jeremy Burgess 25cr.
Geoff Dann 44tl, 44cr, 44bl, 45bc, 45tl, 50bl, 50r, 54bc, 56c, 57tl, 57bl.
Andreas Einsiedel 45tr.
Jacqui Hunt 19tr.
Andrew McRobb 39r.
Karl Shone 17bc.